SAVE THE ANIMALS!

GORILLAS

by J. P. Press

Consultant: Beth Gambro
Reading Specialist, Yorkville, Illinois

BEARPORT PUBLISHING

Minneapolis, Minnesota

Teaching Tips

Before Reading

- Look at the cover of the book. Discuss the picture and the title.
- Ask readers to brainstorm a list of what they already know about gorillas. What can they expect to see in this book?
- Go on a picture walk, looking through the pictures to discuss vocabulary and make predictions about the text.

During Reading

- Read for purpose. Encourage readers to think about how to help gorillas as they are reading.
- Ask readers to look for the details of the book. What are they learning about the problems gorillas are facing?
- If readers encounter an unknown word, ask them to look at the sounds in the word. Then, ask them to look at the rest of the page. Are there any clues to help them understand?

After Reading

- Encourage readers to pick a buddy and reread the book together.
- Ask readers to name one problem from the book gorillas face and one solution. Find the pages that tell about these things.
- Ask readers to write or draw something they learned about saving gorillas.

Credits:
Cover and title page, © Sportactive/iStock; 3, © Eric Isselee/Shutterstock; 5, © Elvis01/Shutterstock; 6–7, © guenterguni/iStock; 9, © Elvis01/Shutterstock; 10, © PeterVrabel/Shutterstock; 12–13, © Nick Greaves/Shutterstock; 14–15, © Brent Stirton /Getty Images; 16–17, © Judith Engbers/iStock; 19, © Syda Productions/Shutterstock; 20–21, © guenterguni/iStock; 22, © Nick Greaves/Shutterstock; 23TL, © PeopleImages/iStock; 23TR, © Wavebreakmedia/iStock; 23BL, © ProStockStudio/Shutterstock; and 23BR, © EdoTealdi/iStock.

Library of Congress Cataloging-in-Publication Data

Names: Press, J. P., 1993- author.
Title: Gorillas / by J.P. Press.
Other titles: Save the animals!
Description: Minneapolis : Bearport Publishing Company, 2022. | Series:
Save the animals! | Includes bibliographical references and index.
Identifiers: LCCN 2022002328 (print) | LCCN 2022002329 (ebook) | ISBN
9781636917931 (library binding) | ISBN 9781636918006 (paperback) | ISBN
9781636918075 (ebook)
Subjects: LCSH: Gorilla--Africa--Juvenile literature.
Classification: LCC QL737.P94 P74 2022 (print) | LCC QL737.P94 (ebook) |
DDC 599.884--dc23/eng/20220119
LC record available at https://lccn.loc.gov/2022002328
LC ebook record available at https://lccn.loc.gov/2022002329

For more information, write to Bearport Publishing, 5357 Penn Avenue South, Minneapolis, MN 55419. Printed in the United States of America.

Contents

A Furry Family

Ooh, ooh, ooh!

Gorillas call to one another in between trees.

But the forest is getting quieter.

Where are the gorillas going?

There are four kinds of gorillas.

All **wild** gorillas live in Africa.

They make their homes in warm, wet forests.

There used to be lots of gorillas in forests.

Now, not many live there.

There are only 300 left of one kind.

Some people **hunt** gorillas.

They eat the animals.

Gorillas are also taken from forests.

They are put in zoos or studied by **scientists**.

Gorilla homes are in danger, too.

People are cutting down forests.

The animals have less space to live.

Some people try to help gorillas.

They stop other people from harming the animals.

People are also keeping trees from being cut down.

They make forests safe for gorillas.

How can you help?

Paper is made from trees.

Recycle your paper!

Then, more trees will be left for gorillas.

Look at those furry faces in the forest.

Together, we can save the gorillas!

Gorillas in the Wild

Where gorillas live

Glossary

hunt to kill animals for food or sport

recycle to turn used, unwanted materials into new, useful materials

scientists people who study nature and the world

wild from places outdoors where animals can live and travel freely

Index

Read More

Best, B. J. *A Gorilla's Nest (Animal Homes).* New York: Cavendish Square, 2019.

Duling, Kaitlyn. *Mountain Gorillas (Blastoff! Readers: Animals of the Mountains).* Minneapolis: Bellwether Media, 2021.

Learn More Online

1. Go to **www.factsurfer.com** or scan the QR code below.
2. Enter "**Save Gorillas**" into the search box.
3. Click on the cover of this book to see a list of websites.

About the Author

J. P. Press likes being outside. Her favorite things to do are running and watching wild animals.